ON

FORCE OF FALLING BODIES

AND

DYNAMICS OF MATTER,

CLASSIFIED WITH PRECISION TO THE MEANING OF DYNAMICAL TERMS.

BY

JOHN W. NYSTROM, C.E.

PHILADELPHIA:
J. B. LIPPINCOTT & CO.
1873.

LIPPINCOTT'S PRESS,
PHILADELPHIA.

FORCE OF FALLING BODIES.

A DISCUSSION on this subject has originated in the *Scientific American*, from a question published in that journal, June 8th, 1872, as follows:

"*Force of Falling Bodies.*—We have a steam-hammer weighing exactly three tons, including piston and rod; the stroke is four feet, and the hammer falls by its own gravity. What will be the force of the blow, making no allowance for friction? What is the formula for the calculation?

"(Signed) "J. E."

The following answers were given:

[*From the Scientific American, July* 6, 1872.]

Force of Falling Bodies.

To J. E., query 12, June 8.—The hammer will strike with a momentum of 160,164.5472 pounds. The formula is

$$\text{the square root of } (4 \times 64.33) = 16.0312 \text{ velocity.}$$

Then

$$4.426 \times 6000 \times 16.0312 = 160164.5472.$$

Or, multiply the fall in feet by 64.33; the square root of the sum is the velocity; and multiply the weight in pounds by 4.426 and that by the velocity, and you have the momentum.

E. E. W., of W. Va.

[*From the Scientific American, July* 13, 1872.]

Force of Falling Bodies.

If J. E., query 18, page 385, last volume, will multiply the weight of any falling body, in pounds, by the height of the fall in feet, he will have the force of the blow in foot pounds. Leaving friction out of the question, the force of the blow of his hammer is precisely equal to the force expended in raising it, namely, 6000 × 4 = 24,000 foot pounds. Converted into heat, this force would be competent to raise the temperature of one pound of water a little more than 31°, thus: 24,000 divided by 772 equals 31.09 units of heat.

W. H. P., of Iowa.

[*From the Scientific American, August* 10, 1872.]

Force of Falling Bodies.

In view of the difference between the two answers to J. E., query 12, June 8, and of my own ideas, somewhat different from either, I would say: The striking force of a moving body, in whatever direction it moves, is its momentum. Its momentum is the joint result of its quantity of matter and its velocity. The ratio of this momentum to that of other moving bodies is compounded of the ratio of its quantity of matter, which is indicated by its weight, and of its velocity at the instant in question. Its momentum, therefore, is not weight any more than it is space or time, and it cannot be expressed by pounds, in the ordinary sense of that word, any more than by feet or by seconds, nor is it expressed by any two of those terms. To obtain a statement of the momentum of a body for the purpose of comparison: Multiply its weight by its velocity—its number of pounds, for instance, by the number of feet it would move in a second if it should proceed for a second at the rate for the instant in question. The velocity of a falling body is continually accelerated, and it increases not as the space fallen over but as the square root (~~query ? Ed.~~) of that space. Therefore, to multiply the weight by the space fallen over, will not give the momentum. The velocity of a falling body at the end of one second of its fall is $32\frac{1}{6}$ feet per second, and it has fallen one-half that distance. It will fall $4\frac{1}{48}$ feet in half a second, and its velocity is then $8\frac{1}{24}$ feet in half a second. The velocity at four feet descent is nearly the same, but more exactly is 16.0312 feet per second. This multiplied by the weight in pounds gives the momentum. The general formula is: The square root of (64.33 multiplied by the distance fallen) = the velocity, and the velocity multiplied by the weight = the momentum. So much for determining the momentum. The extent of change produced by the blow of a hammer has a compound relation to the force of the blow and the ability of that which it strikes to resist. Some obstacles resist in proportion not only to intrinsic power, but also to the time during which they exert their resistance, and their resistance to a blow is less as the velocity of the blow is greater. Such are the different attractive, repulsive, and expansive forces, and such is substantially the case where springs are to be bent and where many forms of cohesion are to be overcome. In such cases, the change produced is as the weight multiplied by the square of the velocity, and in case of a falling body is as the weight multiplied by the distance fallen. Other resistances are independent of time, and are in proportion to the space over which the resistance operates. Such is substantially the case of friction. Here the change is as the momentum of the blow. It is so in the case of bodies resisted by the momentum or inertia of other bodies, or, as in greater or less degree is the case of a body moving through liquids, of the particles of bodies. The case of forging with a hammer presents a compound of both these kinds of resistance, varying in their proportions with the nature of materials, degree of heat, and other considerations.

G. M. T.

[*From the Scientific American, August* 10, 1872.]

Force of Falling Bodies.

To the Editor of the Scientific American:

Since you are publishing a series of articles on "Weight, Pressure, Power, Force," etc., it would be useful to so explain the acting force of a body in motion, its momentum or striking force, that, if such a thing be possible, your readers may understand by what means, by what it is measured, and how determined.

While this is one of the simplest problems in physics, as well as one of the most essentially practical, it is one of those of which the majority of the people are most profoundly ignorant, as is shown by the frequent questions on the subject in your valuable paper, and by the replies, no two of which are alike, and which indicate that the correspondents are hopelessly befogged.

In your number of July 6, page 10, a correspondent—misled by Haswell, probably—estimates the force of the hammer, weighing three tons and falling four feet, at over 160,000 pounds. But what does he mean? What is a pound of force? To what is it equal? What work will it do? He does not say foot pounds, and if he means that, he is wide of the mark in his estimate. A blow cannot be compared with weight or pressure alone.

It should be universally known, if possible, that force is estimated by the measure of the work it is competent to perform, the number of pounds it will raise one foot high. The force which will lift one pound one foot is called a foot pound, and is the unit used to express the amount of a force. Gravitation, being a constant quantity, is a convenient standard, and force measured by the amount of gravitation it will overcome affords a statement quite intelligible to any intelligent person. Next, it should be known that this same one pound, in falling freely one foot, will accumulate the same amount of force, that is, gravity will impart to it in its descent the same amount of force which it took from it in its ascent, and therefore the force of the blow will be just one foot pound; and, if converted into heat, would produce exactly the amount of heat which would be required to lift the one pound one foot high again.

In general, the force with which any falling body will strike is precisely the amount required to lift the same body to the height from which it fell. When, therefore, the weight and height are given, their product is the force of the blow in foot pounds, and in the case of this hammer, would be $6{,}000 \times 4 = 24{,}000$ foot pounds. The force of a "weight of one pound falling two feet" would be $1 \times 2 = 2$ foot pounds, while Haswell's "Engineer's and Mechanic's Pocket Book," page 419, gives it at 11.34 pounds, whatever that may mean.

If the velocity is given, we find the height as follows: Dividing the velocity by $32\frac{1}{6}$ (the velocity acquired in each second) gives the time of fall in seconds, and multiplying the square of the time by $16\frac{1}{12}$, we have the height from which the body must have fallen to acquire the given velocity, which, of course, is also the height to which the body would ascend, if projected upward with the same initial velocity before its force would be expended in overcoming gravitation. Obviously, the force of the blow will be the same, with the same velocity, whether the motion be downward, upward, or horizontal; hence, to find the force with which it is moving, we only require to find the height from which a body must fall to acquire the given velocity, and said height, multiplied by the weight, gives the striking force in foot pounds, or the amount of work the body would perform, the resistance it would overcome, the weight it would lift one foot, or the heat it would produce; and also, what is the same thing, we have the amount of force expended in imparting to the body the given velocity.

The general confusion of ideas upon this subject is probably largely due to the fact that the text-books differ widely, and the majority of them are entirely wrong, as they almost all teach that the striking force is proportional to the velocity, whereas it is, in fact, proportional to the square of the velocity, as is readily shown by the law of falling bodies enunciated in the very same books.

The formula above given is far more simple than the various arbitrary and fantastic ones so often presented by your correspondents, and has the peculiarity of being correct, and consequently consistent with all the laws of motion; and if you will give me space for a few examples, I believe its application will be perfectly plain to your readers. Instead of dividing the velocity by 32.16 and multiplying the square of the quotient by 16.08, we may, of course, obtain the same result by the shorter process of dividing the velocity by 8.02, and squaring the quotient.

1. A one pound ball moves 1000 feet per second; $(1000 \div 8.02)^2 = 15{,}545$. Its force then is 15,545 foot pounds, and as it weighs one pound, if its motion were directly upward it would mount to the height of 15,545 feet, and on returning would acquire in its descent the same velocity of 1000 feet. The force expended, then, in imparting this velocity was equivalent to that required to raise 15,545 pounds one foot.

2. A twenty-four pound ball has a velocity of 50 feet per second; $(50 \div 8.02)^2 \times 24 = 931.44$ foot pounds. If this twenty-four pound weight were a hammer with a stroke of 38.81 feet, it would acquire a velocity of 50 feet, and would strike with a force of $38.81 \times 24 = 931.44$ foot pounds, and this amount of force, in any available form or mode of manifestation, would be sufficient to impart a velocity of 50 feet to a mass of 24 pounds, or to lift 24 pounds 38.81 feet, or to lift or throw one pound 931.44 feet high, or 931.44 pounds one foot high. In these calculations, there is no allowance made for atmospheric resistance.

W. H. Pratt.

Davenport, Iowa.

Neither of those four communications answers the question of Mr. J. E., which is, *What will be the force of the blow?*

They state, how many foot-pounds, how much work, and how much momentum there is in the hammer, neither of which is force, as is required by the question. As I considered it difficult for the inquirer to make out anything from those answers, I volunteered to answer the question in my own way, as follows:

[*From the Scientific American, August* 31, 1872.]

Correspondence.

The Editors are not responsible for the opinions expressed by their Correspondents.

Force of Falling Bodies.

To the Editor of the Scientific American:

The question "With what force does a falling body strike?" has been frequently repeated in the *Scientific American* for the last twenty-five years, and has generally been answered by the batch of dynamical terms used in colleges and styled "scientific." The answers have invariably made the problem more obscure. Each one generally says that "the problem is very simple," and he pretends to understand the subject perfectly. I am one of those pretenders, and propose to answer the question in my own way, reference being made to the accompanying figure.

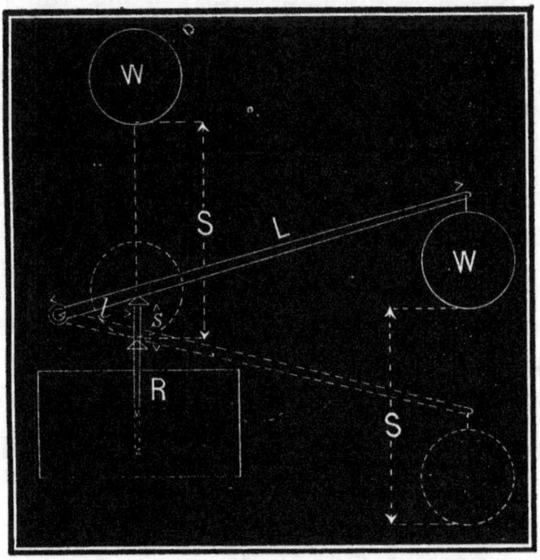

Let us assume the case of driving a nail into a piece of wood by the aid of a lever whose fulcrum is at *C*. The applied force is represented by the weight *W*, acting on the lever *L*. Let *R* denote the force of resistance in the wood, expressed by the same unit of weight as that of *W*, say pounds.

The weight *W*, acts on the long lever *L*, and the resistance *R*, on the short lever *l*. Then

$$R : W = L : l, \text{ and } R = \frac{WL}{l}$$

That is to say, the force of resistance in the wood is to the weight or force W, as the long lever L, is to the short lever l.

Let S represent the vertical height which the weight W moved, and s the distance which the nail was driven into the wood. Then

$$R : W = S : s, \text{ and } R = \frac{WS}{s}.$$

That is to say: the force of resistance in the wood is to the force or weight W, as the height S, is to the distance s.

Now let the same weight W, fall from an equal height S, directly upon the head of the nail, and the latter will be driven into the wood the same distance as by the aid of the lever. Therefore: the force with which the falling body acted upon the nail is to the weight of the falling body as the height of fall is to the distance the nail is driven into the wood. The force of the falling body is equal to its weight multiplied by its height of fall, and the product divided by the distance which the nail is driven into the wood.

JOHN W. NYSTROM.

PHILADELPHIA, PA.

My explanation on force of falling bodies was not well appreciated by an employee of the *Scientific American*, Dr. Vander Weyde, who made the following remarks on the subject:

[*From the Scientific American, September* 14, 1872.]

Correspondence.

The Editors are not responsible for the opinions expressed by their Correspondents.

Force of Falling Bodies.

TO THE EDITOR OF THE SCIENTIFIC AMERICAN:

I see, in an article on page 131 of your paper, that Mr. John W. Nystrom acknowledges himself to be "one of those pretenders" who think that they "understand perfectly the subject" of measuring the force of a falling body by taking, as unit of measurement, the mere weight of matter without motion. I desire here to say to him and to all those interested in the important question of measuring forces, in consideration that force must be distinguished from mere weight, that weight is merely a measure for an amount of matter, for a mere mass, and for nothing else; such weight, of course, is caused by gravitation, and thus can exert pressure, but as long as the weight does not produce motion, there is no force generated; therefore strictly speaking gravitation is no force, notwithstanding the conventional way of speaking of the force of gravitation; however, gravitation can beget force, and only does so in case it is allowed to produce motion. According to the modern conception of force, it is not something immaterial, independent of matter, but absolutely nothing but matter in motion. This motion may be hidden, molecular, when the force manifests itself as heat, electricity, etc., or the motion may be in the masses, when the force is directly measurable by two elements, the mass and the velocity. Accepting the customary symbols for these two different elements, the different degrees of force are expressed by the formulæ $v \times m$ and $v^2 \times m$, which are both correct according to circumstances. In the case of the effect of a blow produced by a falling body, for instance, the driving in of a nail, the identical case represented on page 131, the latter formula, corresponding with the theory of the *vis viva* (see any text-book on mechanics), must be applied. This is the first point in which the formulæ of Mr. Nystrom are faulty, as they are based on the lever, and thus not on the square of the velocity or space, but on the simple velocity : $v \times m$.

The result of this law of the *vis viva* is that, where gravitation increases or decreases, and with it the velocity of the falling body, the force of the blow will increase or decrease as the square of the gravitation, while the weight of the body will only increase or decrease in the simple ratio of the gravitation. Mr. Nystrom's figure and formulæ fail to take any account of this whatsoever.

But let us consider the expressions $v \times m$ and $v^2 \times m$ theoretically. It is evident that they have no value at all as soon as either of the quantities v or m becomes immeasurably small or disappears. Let, for instance, in the function $v \times m$ or $v^2 \times m$,

m become $=0$; then we have $v \times 0 = 0$ and $v^2 \times 0 = 0$, which conform to practical experience, because a blow with a mass equivalent to nothing must necessarily amount to nothing. Let, inversely, v be $=0$, and we have

$$0 \times m = 0 \text{ and } 0^2 \times m = 0,$$

again equivalent to nothing; a mathematical proof that a mere mass without velocity (motion) cannot possibly be reckoned equivalent to any force; and we see here the great mistake, thus far made by the authors of many text-books, in speaking of a force of, say, 100 pounds, or a ton.

The cause of this error is mainly to be found in the fact that a mere weight by its pressure will in some cases produce results similar to that of a force or blow. If, however, we attempt to measure force (matter in motion) by mere weight (matter in rest), we must continually fail and obtain incongruous results, as they are two incomparable quantities. This confounding of an actual force produced by a moving mass with mere weight or pressure produced by a stationary body, is the cause of fifty per cent. of the attempts, continually being made by the half educated, to find perpetual motion.

Now for a few practical illustrations: With a comparatively light hammer, we may easily drive a nail into a brick wall; if we try to do it by mere pressure, we shall crush the nail, or, to take Mr. Nystrom's own illustration, we can drive a nail into a board by the blow produced by dropping the head of a hammer on it from a suitable height, directed by guiding pieces, as in a pile driver; but take a similar nail, place it on the same board, attach the lever of proper length and hang the hammer head at the end of the lever, following practically the figure on page 131, and see if the nail will penetrate at all. If Mr. Nystrom had tried the experiment, he surely would never have taken the trouble to illustrate and publish his explanation.

The blow or percussion gives to a mass a shock, transmitted through it with the same velocity as a wave of sound would travel in that same mass; when the blow is violent and there is somewhere a want of continuity, or lack of strength, which prevents the wave from pursuing its course, its power will be expended there in crushing the material. This is the case in driving a nail. The motion will not be communicated to the board, but the force will be expended in crushing and cutting the fibres of the wood under the nail, so as to allow it to enter, while a weight or pressure placed on the nail will have plenty of time to communicate itself to the whole board. A striking illustration of this may be had when balancing a heavy board on its centre; it is then possible to drive a well-pointed nail with a smart blow deep in the board without moving the latter, while the same nail with a weight on top will scarcely make a mark on its surface, but will move the whole board. A pistol ball may be fired through a door without moving it on its hinges, which latter may be done by the slightest pressure of the finger. Scores of other familiar examples may be adduced, all proving the immense difference between force and mere pressure, and it is only to be wondered at and at the same time deplored that still so much confusion prevails in regard to this all-important subject.

P. H. VANDER WEYDE.

NEW YORK CITY.

The inquirer Mr. J. E., and other readers of the *Scientific American*, will no doubt be well informed on force of falling bodies, after having read Dr. Vander Weyde's philosophy on that subject. When I read the article I could not credit the possibility that Dr. Vander Weyde has ever graduated as doctor of philosophy in any creditable college; which, however, would make no difference to me if he has or not, as long as he keeps himself within his due limits. When an individual bears the title of Doctor or Professor we expect him to issue authority; but when he imposes upon the public with quack philosophy, as Dr. Vander Weyde has done, we have reason to suspect that there may be something wrong about his title, and we are justified in asking the doctor to show his diploma, and in holding him responsible for such offences. With this motive I addressed the editor of the *Scientific American* on the subject, and received the following answer:

[*From the Scientific American, September* 28, 1872.]

Who is Dr. Vander Weyde ?

During the past few weeks, an esteemed correspondent, J. W. Nystrom, Esq., C. E., of Philadelphia, has furnished to our readers several interesting communications, some of which have been answered and criticised by another of our valued correspondents, Dr. P. H. Vander Weyde, of this city. From the tenor of the following letter it would seem that our Philadelphia correspondent is a little suspicious of the respectability of his antagonist. But we can assure him that, in Dr. Vander Weyde, he has a foeman worthy of his lance.

TO THE EDITOR OF THE SCIENTIFIC AMERICAN:

SIR:—Will you be kind enough to inform me, through the *Scientific American*, if Mr. P. H. Vander Weyde, of New York, has a doctor's diploma, and if so, from which college he has received that title? And what kind of a doctor is he?

The answer to these questions will greatly oblige yours very respectfully,

JOHN W. NYSTROM.

1010 SPRUCE STREET, PHILADELPHIA, September 7, 1872.

We would inform our correspondent that Dr. Vander Weyde is a physician of the strictest orthodox sect; that he is an honored graduate of the New York University Medical College, of which John W. Draper, LL.D., is President; that he holds the regular diploma of that institution; that he enjoys the fellowship and esteem of many of our leading physicians and prominent men of science; that he is a native of Holland, where he received a university education; took the degree of Doctor of Philosophy in 1840; was the editor of a scientific periodical; in 1845, at Amsterdam, he received the honorary prize, consisting of the gold medal of the Society of Sciences, for his essays upon natural philosophy.

Dr. Vander Weyde is now a citizen of the United States. From 1859 to 1864, he was Professor of Physics, Higher Mathematics and Mechanics at Cooper Institute in this city. During nearly the same period, he was also Professor of Chemistry in the New York Medical College. From 1864 to 1866 he was

Professor of Industrial Science in Girard College, Philadelphia, Pa. His contributions to the scientific literature of the day have been very extensive, and are widely known.

These are only a few of the items of Dr. Vander Weyde's public record. But they are sufficient, we trust, to satisfy the inquiries of our correspondent, and remove from his mind any adverse prejudices that he may have formed concerning the qualifications of the distinguished gentleman whose public standing he has questioned.

Upon the authority of the *Scientific American* it appears that the said Vander Weyde has really graduated as doctor of philosophy in some college or university in Holland, which, however, I shall not feel convinced of until I have seen his diploma, or am informed of the fact direct from that university.

Holland stands very high in sciences, in fact, on the level with other nations, and its colleges are very strict in their studies and examinations, in which it is doubtful whether Dr. Vander Weyde's philosophy on dynamics of matter could have passed for good.

I cannot enter into any discussions in the *Scientific American* with such a philosopher as Dr. Vander Weyde has proved himself to be, for it is the duty of the university in Holland from which he bears his diploma (if he has any such) to have instructed him in the sciences in which he has proven himself deficient.

As soon as I can find out the name of that university, I shall endeavor to have the subject attended to, and if that university cannot endorse and sustain its doctor's philosophy, he ought to be called back and made to study over again, or be requested to return his doctor's diploma.

I have learned that he has been professor of natural philosophy in Girard College, Philadelphia; and if he has advanced such philosophy there as that he has done on force of falling bodies, his students are simply deceived.

If Dr. Vander Weyde had limited himself within the confusion which really exists in the want of precision to the meaning of dynamical terms, I should have had forbearance with him; but he dictates a doctrine which is contrary to established facts in physics.

It may be considered that I am rather severe upon Dr. V. W., but there are so many of this kind of philosophers that it will do no harm to tell the truth to one of them every now and then; and what has been said about Dr. V. W. is equally applicable to all the rest of the high authorities who have invariably attacked me in the style of quack der Weyde. When my ideas differ from what is written in their books, they blindly suppose that I am wrong, and they attack me with irrelevant philosophy, by which the public has been juggled, and I have been embarrassed all my life by quack opinions of high authorities. In many cases I have doubted whether the high authorities have themselves believed in their own statements, but they evidently expected me to have faith in their profound reasonings of perfect nonsense.

The object of this writing is, however, not to attack Dr. V. W., nor to prove that I am right and that he is wrong, all of which is of secondary importance to me and to the public; but my principal object is to call the serious attention of scientific institutions to the confusion in dynamics, and to the annoyance which that confusion causes to the public, and between individuals.

For the last ten years I have had repeated discussions on the confusion of dynamical terms, in which my opponents have not been able to sustain themselves. Some of these discussions are published in the *Journal of the Franklin Institute*, and also in the *Scientific American* for the year 1865. The high authorities generally maintain that work is independent of time!!! which is contrary to the opinion of manufacturers in regard to the eight or ten hours' struggle for a day's work.

We have no good text-books on dynamics, and, even in colleges, the subject is confused, not so much in substance as in terms, and the result is that graduated students, even from the same college, differ as to which is which in dynamics. The confusion is, however, acknowledged by high authorities, who have not been able to sustain themselves against me in open contests, and, when such authorities are professors in colleges, the public cannot thus receive proper information, and the subject remains obscure, as proven in the case before us. Under such circumstances I have a right to claim the issue of authority, which is as follows:

CLASSIFICATION OF DYNAMICAL TERMS.

I do not acknowledge the existence of more than one kind of force in physics, and that is, that action which can be expressed simply by weight, without regard to motion, time, power, or work. Force is derived from a great variety of sources; but when it is simply force, it can always be expressed by weight.

Force, motion, and time are simple physical elements. Space, power, and work are functions of those elements.

Elements.			*Functions.*	
Force	$= F$	1.	Space $S = VT$	4.
Motion	$= V$	2.	Power $P = FV$	5.
Time	$= T$	3.	Work $K = FVT$	6.

The weight of a body is the force of attraction between that body and the earth; and as the force of attraction varies inversely as the square of the distance between the centres of the attracting bodies, the weight of a defined body is not a constant quantity.

Mass means the real quantity of matter in a defined body, and it is a constant quantity which cannot be increased or diminished by force. Mass is generally denoted by the letter M, and is one of the four elements which constitute the dynamics of matter.

The relation between these four elements is well known and accepted as an established fact, namely,

$$M : F = T : V \quad . \; . \; . \; . \; . \; . \; . \; . \quad 7.$$

This is the fundamental analogy in the dynamics of matter, of which the two functions,

Momentum of motion, $MV = FT$, momentum of time, 8.

The function MV has been termed *force*, which is a great error, for it only denotes the product of the force and time consumed in giving the mass M the velocity V, and which is equal to the product of another force and time required in bringing the mass from motion to rest. There is no relation between the two forces, which are entirely governed by their respective times of action.

The function 6 expresses the

$$\text{Work } K = FVT \quad . \quad . \quad . \quad . \quad . \quad . \quad . \quad 6.$$

Multiply the momentums function 8, by the velocity V, and we have the

$$\text{Work } K = MV^2 = FVT \quad . \quad . \quad . \quad . \quad . \quad 9.$$

We see here that the function MV^2 means work, but it has been termed *force* and *vis viva*, which are also erroneous. MV^2 only denotes the product of force, motion, and time, which is the work consumed in giving the mass M the velocity V, and which is equal to the product of another force, motion, and time, or the work required or executed in bringing the moving mass to rest.

The function 4 expresses the

$$\text{Space } S = VT \quad . \quad . \quad . \quad . \quad . \quad . \quad . \quad . \quad 4,$$

which, inserted in function 9, will be the

$$\text{Work } K = MV^2 = FS \quad . \quad . \quad . \quad . \quad . \quad . \quad 10,$$

in which the function MV^2 denotes the product of the force and space in which the mass M attained the velocity V, and which is equal to the product of another force and space in which the moving body is brought to rest, and which is the same as the primitive function FVT.

None of these functions should be called force.

The present confusion in dynamics consists in that the functions are called force, for which we can make neither head nor tail of it.

Dynamics is now in the same condition as geometry would be if there was no distinction between *line*, *surface*, and *solid*, and it will remain in that confusion as long as the colleges teach that MV and MV^2 are forces.

The erroneous idea that MV and MV^2 are forces has stuck so tightly into the heads of philosophers as to cause an epidemic in dynamics.

Dr. Vander Weyde says: "Accepting the customary symbols for these two different elements, the different degrees of force are expressed by the formulæ $V \times M$ and $V^2 \times M$, which are both correct according to circumstances." I would ask Dr. V. W. if the circumstances depend upon the weather.

The *Scientific American* of the 22d of June, 1872, gives the following ideas of dynamics:

Weight, Pressure, Force, Power, Work.

The fact that the above words are often confounded together, for the simple reason that their true meaning is not well understood, has been the cause of many fruitless attempts at mechanical inventions and improvements. Most searchers for perpetual motion make no distinction between pressure and force, and are under the delusion that mere pressure can produce work, and we have seen writers on mechanics and we have even heard lecturers on scientific subjects speak of a force of, say, two tons weight. Weight alone is not force, neither is pressure equivalent to work; and it may therefore be useful to attempt some clear definitions of the above terms, in order to protect inventive minds against mistakes in mechanical reasoning.

Weight is simply the measure of an amount of matter referred to a certain standard accepted as a unit. This unit may be a gramme, a pound, a ton, or our whole earth, which the astronomers use; but, in either case, it conveys to the mind nothing but the conception of an inert mass, or a certain amount of matter, for the determination of which gravitation gives us the means of measuring and comparing. Therefore we may say: To have "a mass of two tons," but not "a force of two tons."

Pressure is a result of this gravitation, and a mass of two tons will exert a pressure of two tons; in this way we may estimate the effect of a spring, hydraulic press, or other similar contrivance, by saying its pressure (not its power) is equal to two tons, meaning thereby that it has the effect, on the material to be pressed, as if two tons weight were placed upon it; but we have in pressure neither force nor power. These conceptions of the latter require other elements, as we shall soon see.

Force is matter in motion, nothing more, nothing less; the abstract idea of force without matter is a nonentity. All the modern discoveries in science tend to prove this more and more plainly. Without matter, force would have no existence, but it may be hidden in matter as molecular invisible motion in the form of heat, electricity, etc. The steam engine, electro-magnetic engine, etc., are there to prove how this molecular motion, or hidden force, may be changed into visible force or motion of matter. Inversely, the caloric friction machine changes motion into heat; the ordinary and also the Holtz electric machine change motion into electricity. In any case, we are driven to the conclusion that all force proceeds from motion of matter, and is finally resolved into motion of matter, either of masses, or into molecular motion, generating one of the so-called imponderable forces.

Chemistry has proved since the last century that the amount of matter in the universe is a constant invariable quantity, and that we cannot create or destroy a single material atom, but can only change its form solid to liquid, or gaseous, or *vice versâ*. So the modern philosophy of mechanics proves that the amount of force (that is, motion of matter) in the universe is a constant quantity, and that we cannot create or destroy the slightest amount of this force, but can only change it from mass motion to molecular motion, that is, heat, electricity, etc., or *vice versâ*.

The measure of force is thus the product of the mass with the distance through which it moves; and as the unit of measure of ordinary masses is the pound, and of distances, the foot, we have adopted the foot-pound as the standard unit of force, meaning "one pound *lifted against gravitation* one foot," not "one pound moved one foot," as we have seen and heard it stated, which of course gave rise to the most absurd calculations in regard to the immense power obtained to drive a steamship or railroad train.

If one pound weight is raised one foot, one unit of force is expended; if, inversely, we cause one pound to descend one foot, we obtain a unit of force back, and may transform this into other mass motion, or into molecular motion. We may cause this mass of one pound to be raised slowly if we have little power to apply, or rapidly if we have greater power; and, inversely, we may cause it to descend slowly, as is done in the weight of a clock, and spend itself gradually during a long period of time, producing slight effects throughout that time; or we may cause it to descend quickly, as is the case with the blow of a hammer, and spend itself during a very short period of time, almost instantaneous, producing a powerful effect for that short time. So the driving in of a nail, which often the pressure of a ton weight would not accomplish, the blow of a hammer of one pound, lasting a small fraction of a second, will accomplish easily. This remark points out forcibly the difference between the weight of masses at rest and of masses in motion; in other words, the immense difference between mere pressure and force.

We may understand from the *Scientific American* that modern discoverers have undiscovered the discoveries of Sir Isaac Newton.

The law of *force* of universal attraction, which was established by Newton, is as follows: *The force of attraction between any two masses is proportionate to the product of the masses, and inversely as the square of their distance apart.*

There is no motion in this definition of force, and it makes no difference whether either one or both the masses are at rest or in motion, the *force* will always hold good with the definition. Call $F =$ force of attraction, between the masses M and m, and $D =$ their distance apart.

$$\text{Then } F = \frac{Mm}{D^2} \quad . \quad . \quad . \quad . \quad . \quad . \quad 11.$$

$$FD^2 = Mm \quad . \quad . \quad . \quad . \quad . \quad . \quad 12.$$

$$F : M = m : D^2 \quad . \quad . \quad . \quad . \quad . \quad . \quad 13.$$

Call $V =$ the mean velocity in the time T, in which the masses M and m would be drawn together by their own attraction, and we have $D = VT$ and $D^2 = V^2T^2$.

$$\text{The work } K = FVT = \frac{Mm}{VT} \quad . \quad . \quad 14.$$

The object of this treatise does not involve the development of this very interesting subject, which must be deferred to a more appropriate occasion.

In the application of the physical elements to practice, we must assume units of measures for each of them, namely:

$F =$ force in pounds, acting on the mass M.

$V =$ velocity in feet per second.

$T =$ time in seconds, in which the force F acts on the mass M, and produces the velocity V.

$W =$ weight in pounds of a body, or of the mass M.

When the body falls freely under the action of gravity, then $F = W$.

$M =$ mass, which is proportionate to weight when compared in one or the same locality.

The mass of a body is said to be its weight divided by the acceleratrix g, or 32.17, for which the unit of mass should be a quantity of matter weighing 32.17 pounds.

There is no name adopted for this unit of mass, the want of which makes the subject obscure. If we are told that the weight of a body is 20, we naturally ask what twenty? and we cannot conceive the magnitude of the weight until we know the name and magnitude of its unit. Although we may know that thirty-two pounds of matter is a unit of mass, it does not make the clear impression as if the unit had a name, for which reason I would propose to christen this unit with the name *mat.*, from the word matter.

Then, the unit *mat.* means a quantity of matter weighing 32.17 pounds, and there will be 69.63 mats in a ton weight of 2240 pounds. Now, we can apply our reasonings to practice.

The actual force F in pounds of attraction between any two masses M and m, expressed in mats, and $D =$ their distance apart in feet, will be closely approximated by the formula—

$$F = \frac{M\,m}{5274460 D^2}. \quad . \quad . \quad . \quad . \quad 15.$$

ILLUSTRATION OF DYNAMICS OF MATTER.

Let a force F be applied on a mass M at rest, but free to move; the mass will then be set in motion with an accelerated velocity as long as the force acts. If the force F is constant, the acceleration of velocity will also be constant, and the velocity V, attained in the time T, will be from the fundamental analogy of dynamics of matter 7.

$$V = \frac{FT}{M} \quad . \quad . \quad . \quad . \quad 16.$$

That is to say, the force F acting upon the mass M, in the time T, will generate the velocity V.

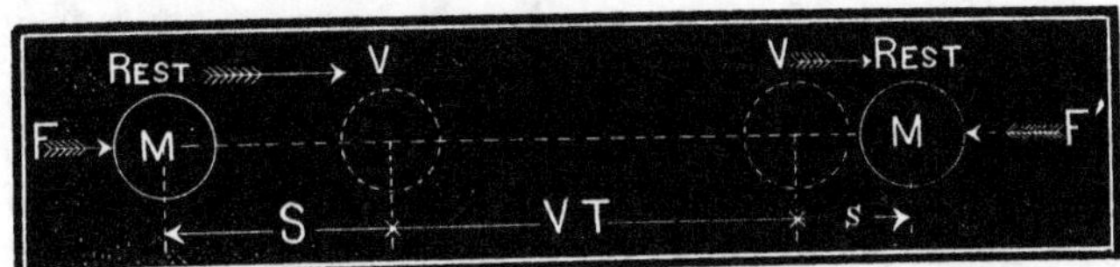

Example.—A force $F = 8$ pounds, is acting upon the mass $M = 4$ mats, for a time of $T = 3$ seconds. Required the velocity V.

$$V = \frac{8 \times 3}{4} = 6 \text{ feet per second.}$$

In the functions 4, 5, and 6, V means the mean velocity in the time T or space S, and when V means the actual velocity of a moving body accelerated from rest to V, or retarded from V to rest, the mean velocity will be ½ V. Then we have the real work expended on the moving mass, or generated in bringing the mass to rest, to be,

$$K = \frac{MV^2}{2} = \frac{FVT}{2} \quad . \quad . \quad . \quad . \quad 17.$$

The space $S = \frac{VT}{2}$, which inserted in formula 17 will be the

$$\text{Work } K = \frac{MV^2}{2} = FS \quad . \quad . \quad . \quad 18.$$

Referring again to the illustration, the force F acted on the mass M only in the space S, after which the body will continue with uniform velocity and generate the space VT, when another force F', independent of F, is applied in opposite direction to stop the motion of the mass M, which is accomplished in the space s, independent of the first space S, but the work consumed or generated will be alike in both cases, namely,

$$K = \frac{MV^2}{2} = FS = F's \quad . \quad . \quad . \quad 19.$$

$$\text{Of which } F : F' = s : S \quad . \quad . \quad . \quad 20.$$

In the case of a falling body, the force F is equal to the weight of the body, and when it strikes a blow, the mean force of resistance is F', acting in the space s, but as the force F or weight of the body is constant even through the space s, the falling body is acted upon by two opposite forces F and F', for which the space S of the fall must also include the space s of the blow.

Then we have the force of a falling body equal to its weight multiplied by the whole height of fall, and the product divided by the space of the blow.

When the force of resistance is irregular, the mean force will still be as the above rule.

In the case of the problem of driving a nail into a piece of wood, as before illustrated, whether accomplished by the aid of a lever or by the weight falling the same space directly upon the nail, the rule will hold good in both cases.

In the case of a steam hammer weighing $W = 3$ tons, and falling $S = 4$ feet, including the compression of the iron by the blow, say, $s = 0.1$ of a foot, required the force of the blow.

$$F' = \frac{WS}{s} = \frac{3 \times 4}{0.1} = 120 \text{ tons.}$$

This is the answer to the original question of Mr. J. E.

The force of resistance F' is not uniform in the space s of compression of the iron forged. It is generally smallest when the hammer first touches the iron, and greatest at the moment the hammer stops, but the mean force throughout the compression s, will be according to the formula.

The *Scientific American* attacked me on the subject of dynamics some seven years ago, as follows:

[*From the Scientific American, June* 29, 1865.]

Work and Power.

In the pages of the *Journal of the Franklin Institute*, a discussion is going on between De Volson Wood, Professor of Civil Engineering in the University of Michigan, and J. W. Nystrom, Acting Chief Engineer, U. S. N., on the subject of work, force and power. The main purpose of Mr. Nystrom seems to be to deny the position that work is independent of time, and he succeeds in involving the question in considerable confusion. The facts of the case are simple and plain enough.

Work is the overcoming of mechanical resistance of any kind, either by raising a weight, dragging a body along, turning off a shaving, or by any other action. The question whether it is independent of the time depends entirely upon the meaning of the language employed. A foot-pound of work is the raising of one pound of matter one foot in vertical height, and this foot-pound is precisely the same quantity whether one second or one thousand years be consumed in the operation.

We may say that a machine is doing the work of raising one foot at the rate of one inch per second; then the work done by the machine will depend upon the time that it is in operation; it will take it twelve seconds to do one foot-pound of work, and twenty-four seconds to do two foot-pounds.

In this case, however, we have attached to the word "work" a meaning for which the word "power" is employed by the standard writers on mechanical philosophy. To keep our ideas clear, it is better to regard the machine as exerting a power of one inch-pound per second, and to confine the word "work" to the aggregate resistance overcome.

One writer argues that 2 and 2 do not always make 4, sometimes making 22. By analogous tricks of language we may confuse our minds in regard to any problem whatever; but a more useful aim of discussion is to free our minds from confusion, and to accomplish this one of the most important steps is to use words always in their exact signification.

Regarding work as the overcoming of physical resistance, it is plain that the aggregate amount of any given quantity is independent of the time required for its performance.

There is probably no higher authority on the philosophy of mechanics than Arthur Morin, and from his "Leçons de Mécanique Pratique," translated by Bennett, we take the following extract:

"*The Idea of Work is Independent of Time.*—We see from what precedes that in the measure of work we have only regarded the effort exerted, and the space described in the direction peculiar to this effort. It is, therefore, independent of time.

"Thus, in raising goods the effect is not measured by the duration of labor, but by the product of the load into the height of its elevation."

The editor of the *Scientific American* declined at first to publish my reply to the above criticism, and he wrote me a letter advising me to consult some authority on the subject, indicating that my views on dynamics were all visionary; whereupon I called at the office of the editor and insisted upon it that my reply should be published, which was at last consented to, and which is as follows:

[*From the Scientific American, September* 9, 1865.]

Force, Power, and Work.

(For the Scientific American.)

FORCE is a mutual tendency of bodies to attract or repel each other. Its physical constitution is not yet known. We only know its action, which is recognized as pressure and measured by weight. The unit of weight being assumed from the attraction of the earth upon a determined volume of any specific substance; for example, the force of attraction between the earth and 27.7 cubic inches of distilled water, at the temperature of 39.8° Fahr., in an atmosphere balancing 30 inches of mercury, at the level of the sea, which is called one pound avoirdupois. Force is the first element of Power and Work, and can be likened to length, which is a primary element in geometry. Force will here be denoted by the letter F, expressed in pounds.

VELOCITY is the second element of Power and Work, and may be likened to breadth in geometry. It is that continuous change of position recognized as motion, and is here denoted by the letter V, expressed in feet per second. Velocity is a simple element, although it appears to be dependent on time and space, but the space is divided by the time, and therefore both relieved from the velocity.

TIME is the third element of work, and may be likened to thickness in geometry. It implies a continuous action recognized as duration. Time is here denoted by the letter T, expressed in seconds.

POWER is a function of the two first elements—force F, and velocity V—as area in geometry is a function of length and breadth. Power is here denoted by $P=FV$, which means that the power P, is the product of the force F, multiplied by the velocity V. The power so obtained is expressed in foot-pounds, and called dynamic effect, of which there are 550 in a horse-power; or if the velocity is measured in feet per minute, there will be 33,000 foot-pounds in a horse-power. Power is independent of space and time, but it has often been confounded with work, which essentially depends on time and space.

SPACE is a function of the second and third elements—velocity V, and time T—and may be likened to a cross section of a solid, which is a function of breadth and thickness. Space is here denoted by $S=VT$, which means that the space S, is the product of the velocity V, and the time T, expressed in linear feet.

WORK is a function of the three elements,—force F, velocity V, and time T. It may be likened to a solid in geometry, which has the three dimensions,—length, breadth, and thickness. Work is here denoted by $K=FVT$, which means that the work K, is the product obtained by multiplying together the three elements—force F, velocity V, and time T.

Work may also be denoted by $K=FS$, or the product of the force F multiplied by the space S, where it appears as if the work was independent of time, but the time is included in the space $S=VT$.

Work may also be denoted by $K=PT$, which means the power P, multiplied by the time T. Either of the three cases expresses the work in foot-pounds.

Force, velocity, and time are simple physical elements.

Power, space, and work are functions or products of those elements.

The *Scientific American* is read by most mechanics in this country, and it may be further said that that journal is met with in most parts of the world. It evinces a habitual and sincere desire to furnish its readers with correct and instructive articles on scientific subjects, in consideration of which it would be a neglect of duty on my part to pass over in silence its articles on "Work and Power," published on page 71 of the present volume. In that article you say, "The main purpose of Mr. Nystrom seems to be to deny the position that work is independent of time, and he has succeeded in involving the question in considerable confusion." And you think "the facts of the case are simple and plain enough." You proceed to give an antithetical description of what work is, and say, "In this case, however, we have attached to the word *work* a meaning for which the word *power* is employed by the standard writers on mechanical philosophy." Now you will allow me to remark that, in this expression, you have, together with the standard writers, confounded *work* with *power*. You have thus not followed your own good advice, namely, "to free our minds from confusion" by taking "most important steps to use words always in their exact signification." You then go on to say, "Regarding work as the overcoming of physical resistance, it is plain that the aggregate amount of any given quantity is independent of the time required for its performance." Do you not here convey the idea that *work is independent of what it requires*, namely, the *time?* You evidently mean to say that a given quantity of work may be performed in any desired length of time; but you do not seem to conceive that the work is dependent on whatever time is required for its completion.

Referring to a geometrical figure, you may be able to comprehend the position of your argument about work, which is substantially this—*the cubic content of a pancake is independent of the thickness required to make it up!* You say, "The question whether it (work) is independent of the time depends entirely upon the meaning of the language employed." To this I respectfully object, inasmuch as I recognize only one meaning in the language I have employed; and, indeed, the entire controversy upon this subject appears to have sprung from a rejection or misappreciation of the specific meaning I have struggled to establish to the terms *force*, *power*, and *work*. But supposing, for the sake of argument, that my general language is not sufficiently clear, if you can read my algebraical formulas you will not misunderstand me. Why, therefore, do you not exemplify your argument upon my formulas and thereby show its effect in practice? To say that "work is independent of time," is to say that work is dependent of no time, or that any amount of work may be performed in no time—a proposition which is not yet realized.

Work, as before stated, is the product of the three elements,—force, velocity, and time. For a given quantity of work, either one or two of these elements can vary *ad libitum*, but only at the expense of the remaining two or one. Work is thus not confined to any specific relation or ratio to either of those elements, but independent of either one of them it ceases to be work.

I am well aware that the standard authors have to this day considered work independent of time, and they have also confounded force, power, and work with each other, so that we are yet thrown upon our individual authority to decide which is right. Thus, when you and Professor Wood cannot defend your position, it may be very convenient to assert that I use "cant phrases" when I exemplify your arguments, and that "by analogous tricks of language we may confuse our minds in regard to any problem whatever." I nevertheless trust that I have given both perspicuity and precision to my language and meaning in this article, and sincerely hope that it may be tributary to "the consummation—devoutly to be wished for"—of reducing to certainty and system the future reasonings of the scientific world on this subject.

JOHN W. NYSTROM.

[*From the Scientific American, September* 9, 1865.]

Nystrom on Work and Power.

We have a kindly feeling towards Mr. Nystrom, having received from him several valuable contributions. It has seemed to us, however, that his method of explaining the difference between work and power was calculated rather to confuse than to elucidate the subject. In reply, he forwards us a communication containing his explanation, with a request that we would lay it before our readers and let them judge for themselves. We comply with his request with pleasure, and the communication will be found on another page.

The raising of one pound of matter one foot in vertical height is one foot-pound of "work." The raising of 33,000 pounds one foot is 33,000 foot-pounds of work, whether one minute or one hundred years be consumed in the operation.

The power—either of a steam-engine, waterfall, or animal—that can raise 33,000 pounds one foot in each minute of time is one-horse power; the power that can raise 33,000 pounds in half a minute is two-horse power; and the power that can raise 33,000 pounds in one-tenth of each minute is ten-horse power.

Morin and other writers, therefore, say that the idea of work is independent of time, but that time is an element in the measure of power. It seems to us that these writers are correct. It seems to us, also, that the matter is extremely plain and simple.

Mr. Nystrom, on the other hand, while accepting, if we understand him, the above illustrations of both work and power, denies that work is independent of time, or that time is an element of power, and asserts that the subject is not generally understood even by educated engineers. We have criticised his arguments upon it as calculated rather to confuse than elucidate it. From this criticism he wishes to appeal to the judgment of our readers,—an appeal in which we cheerfully concur.

I do not doubt the kindly feeling of the editors of the *Scientific American* towards me, which has been proven in their very kind advice; and I am convinced that they have no improper motives in attacking me on dynamics; but the misfortune is that they do not understand the subject.

In regard to work being dependent or independent of time, we have only to refer to the operation of a steam-engine of a definite power. Whatever time the engine is running, say one hour or one year, the power is constant, independent of time;

but the work executed by that engine depends entirely upon the time of operation.

The best joke of all is that the editors of the *Scientific American* are not ashamed of themselves for undertaking to give me advice on the subject. I have lately received a letter, signed Munn & Co., praising Dr. Vander Weyde and advising me to drop the subject, evidently with the intention of frightening me.

The publication of this pamphlet is the result of their last advice. I have heretofore never entertained the slightest idea of giving advice to the editors of the *Scientific American*, but now I feel disposed to do so, namely:

I will hereby advise the editors of the *Scientific American* to inform themselves well on the subject of dynamics, before they give me advice and attack me on that subject.

I am informed on good authority that Dr. Vander Weyde is engaged for the *Scientific American* to write on scientific subjects or as an acting editor, and he writes articles which are published under the head of correspondence, stating that "the editors are not responsible for the opinions expressed by their correspondents."

It will be noticed in this pamphlet that the *Scientific American*, as well as myself, is attempting to clear up the subject of dynamics in regard to definite meanings of its terms, and we are of different opinions about the true meaning of the term force. In the year 1865 I submitted a communication on the subject of dynamical terms to the National Academy of Sciences, which met at Washington that year; but the papers were handed back to me immediately, with information that "the subject is all clear."

In conclusion, I most respectfully invite the attention of scientific institutions to my classifications of dynamical terms, that a decision may be given, whether it be adopted or rejected. A prompt action on this subject, in aid of establishing precision to the meaning of dynamical terms, would render a great service to the public.

www.ingramcontent.com/pod-product-compliance
Lightning Source LLC
LaVergne TN
LVHW011135110826
845150LV00008B/2362
* 9 7 8 1 4 1 8 1 9 4 6 0 4 *